COUNTER SNIPER GUIDE

Compiled and originally published under auspices of the U.S. Army Marksmanship Training Unit, Fort Benning, Georgia

PALADIN PRESS
BOULDER, COLORADO

Counter Sniper Guide
Compiled and originally published under the auspices
of the U.S. Army Marksmanship Training Unit,
Fort Benning, Georgia

ISBN 0-87364-069-1
Printed in the United States of America

Published by Paladin Press, a division of
Paladin Enterprises, Inc., P.O. Box 1307,
Boulder, Colorado 80306, USA.
(303) 443-7250

Direct inquiries and/or orders to the above address.

FOREWORD

With the increase in civil disorders, the term sniper has come into common usage (particularly in the press) which is in general, erroneously used in that the term is commonly applied to any person who fires at a specific area or person with any type of firearm. Webster defines a SNIPER as "a sharpshooter concealed to harass the enemy by picking off individual members, usually at long range, and with a telescope-sight equipped rifle."

Regardless of what we may call him, the individual who is shooting at police, firemen, soldiers or citizens is certainly dangerous. In order to counteract, we must employ a trained individual whose knowledge and skill fall within the dictionary description of a sniper, whom we shall refer to throughout the manual as a COUNTERSNIPER.

This manual provides general basic information, which we hope will be of assistance to those concerned in the selection of equipment, training, and employment of the countersniper.

The contents of this manual pertaining to the selection of equipment are presented in the sequence that it is felt should be followed in determining the end item best suited to the needs of the user.

TABLE OF CONTENTS

ILLUSTRATIONS

CHAPTER 1

AMMUNITION

1. We believe that, in most instances, the countersniper will not be required to engage targets beyond 300 yards. In fact, it is our belief that the majority of the targets will probably fall within a span of 100 to 200 yards, particularly in built up areas. Infrequently, however, perhaps in the countryside, a target may be engaged at distances up to perhaps 500 to 600 yards.

2. Of the several factors that must be considered in the selection of countesniper equipment (i.e., ammunition, weapons, and scopes), accuracy is perhaps the most critical, because on many occasions, the target presented will be small, and in addition possibly partially obscured or blending into the background. Further, innocent people may be located in the immediate vicinity of the target and the countersniper must avoid injuring bystanders.

3. The ammunition to be used is the first item to be considered in terms of your objective or mission. Once you determine the ammunition that best suits your requirement, you may then consider the weapon and scope that will enable the countersniper to place the round in the target. The type of bullet is a prime factor to be considered. Soft nose, hollow point or other types of commerical hunting bullets would certainly be the most effective, though not necessarily the most accurate. However, public officials are continually confronted with public opinion and "Humanitarians" who may class this type of bullet as barbaric; and therefore, in some instances, the public officials concerned may be forced to restrict the employing agency to the use of military-type bullets of full jacketed configuration.

4. Noise level and recoil are certainly factors to be considered inasmuch as the average shooter is able to obtain better performance from a rifle of low recoil rather than one of medium or heavy recoil. It is also advantageous to have a low level of noise for the following reasons:

a. We do not want to alert the sniper or snipers to the fact that they are being fired upon.

b. Eliminate, or certainly reduce, the possibility of panic on the part of the general public, which often accompanies gunfire.

c. Reduce the shock effect on the ears of the countersniper who will perform more precisely with a low level of muzzle blast.

5. For distances up to 300 yards, we recommend the Caliber .222 Remington cartridge. Commercial loadings by our American ammunition companies will usually shoot ten shot groups of one inch or less at 100 yards. Many ammunition lots will, in fact, shoot groups of .6 to .7 of one inch at this distance. The recoil is mild, noise level low and the cost of ammunition relatively inexpensive. The performances capabilities of this cartridge are shown in Annex A. Please note that the rifle was sighted to hit "point of aim" at 250 yards, and all groups fired at other ranges are in relation to this same aiming point, indicating that the bullet will strike approximately four and a quarter inches above this aiming point at 100 yards, a little over five inches above at 200 yards and four inches below at 300 yards.

At ranges greater than 300 yards the .222 cartridge is considered impractical, due to the loss of velocity, which results in drastic changes in the trajectory curve that would negate the possibility of disabling wounds. For instance, using the same sight setting explained above, we would find the strike of the bullet two feet below the aiming point at 400 yards and five feet below at 500 yards. We acknowledge that a full-sized man could be hit at these greater distances by careful estimation of the range and the proper "hold over"; however, the countersniper ordinarily cannot depend on getting a shot at a fully exposed target and must therefore have the capability of hitting a much smaller target.

6. For the greater ranges we recommend a cartridge of the .30 caliber class using either the .308 Winchester (7.62mm NATO) or 30-06. Bullets of all types commercial loaded can be obtained in either caliber cartridge. Highly accurate MATCH ammunition is available from commericial sources as well as from government contractors. These cartridges have sufficient size, energy, velocity and bullet weight to produce satisfactory disabling wounds at greater ranges.

7. Many people will undoubtedly wonder why we have not recommended other calibers such as the 22-250 Remington, 220 Swift, 243 Winchester, 6mm Remington .270, .284, etc. In many cases they are all, entirely suitable insofar as accuracy and flat trajectory are concerned, however, they are not available with full metal jacket bullets. If the countersniper possesses an accurate Varmint rifle and can obtain or handload bullet types suitable to the area and conditions, by all means use it, rather than go to the additional expense of special rifles. If however, if you are making an initial investment, we suggest that you give our recommendations in chapter 2 careful consideration.

CHAPTER 2

RIFLES

1. The rifle used by the countersniper must be as accurate as it is reasonably possible to obtain. The USAMTU shop builds rifles that will give an average extreme spread of one and a half inches at 300 meters (328 yards), but these rifles are constructed by the world's most experienced gunsmiths and the cost of the barrel alone exceeds the cost of a rather good commercial rifle. Rifles meeting the above mentioned accuracy criteria are impractical from the standpoint of cost and availability. From a realistic point of view, insofar as accuracy requirements are concerned, we recommend a rifle that will constantly shoot into two minutes of angle or in terms of measurement, two inches at 100 yards, six inches at 300 yards, etc. To be acceptable, the rifle must consistently shoot these size groups or smaller.

NOTE: Some users may feel that a requirement exists for a weapon with a rapid fire capability. However, to our knowledge, the only semi-automatic rifle that has an acceptable accuracy insofar as sniping is concerned, is the US Army XM-21 sniper rifle. At this time, these rifles are only available to Military Police Units.

2. The manually operated bolt action rifle is the most accurate of all of the commercial versions available and is by far the easiest type to modify to produce the quality weapon required for sniping. The heavy or so called "Varmint" weight barrel is an absolute necessity in order to ensure an accurate and stable rifle. The relative merits of the heavy barrel versus the standard of "Sporter" barrel, insofar as stability is concerned, can be illustrated by comparing the movement of the pendulum of various clocks. The movement of the pendulum of the typical coo coo clock is very fast and changes direction at a rapid rate, whereas the heavy pendulum of the grandfather clock swings slowly back and forth. Let us assume that a set of sights are attached to both types. Immediately it becomes apparent which one would allow a more precise alignment and timed release of a shot. A light hunting rifle has its place when a long carry or hard climb up a mountain is required, but it is too inefficient to be used as a countersniper weapon.

3. The countersniper rifle must be capable of holding a ZERO. For example, it must shoot in the same place in relation to the aiming point every day regardless of changes in weather, temperature, humidity level, etc. Therefore, the modifications listed below should be made to the rifle selected in order to obtain the accuracy and consistency that is required.

a. Epoxy Impregnation of the Stock. In order to prevent warpage and principally to increase the tensile strength of the stock, impregnation of the wood of the stock with an epoxy is recommended. This is rather impractical for a small organization with only one or two rifles, but for the larger departments with a number of weapons, it is quite advantageous. A brief description of this process can be found at the rear of the manual. (Annex B.)

b. Free Floating of Barrel. Wood should be removed from the barrel channel in the stock, until the barrel is entirely free of the stock. This is called "free floating" and assures that the barrel will be allowed to vibrate freely and uniformly on each shot. The clearance between the barrel and stock should be a minimum of 1/16 inch. It has been customary in the past to use a dollar bill to ensure proper clearance, by sliding it up to the front of the receiver between the stock and barrel. This is not sufficient clearance, as many barrels will vibrate several thousandths of an inch and strike the stock in an inconsistent manner causing inaccuracy and shifting of the center of the shot group. After the wood has been removed, the barrel channel of the stock should be coated with a good grade of varnish (bar top varnish is excellent) to close the pores of the wood and prevent changes in the dimensions of the wood that would result from the moisture that would be either absorbed or conversely lost due to fluctuations both in the humidity level and temperature.

c. Glass Bedding. In order to obtain the desired consistency we must join the metal parts of the mechanism to the stock by means of glass bedding. "Bedding" is a term used to describe the matching fit between the metal parts and wood of the stock. Glass bedding is the use of either a fiberglass or a plastic type material to obtain a perfectly matching fit between the metal and wood. The material is very similar to that used in the manufacture of boat hulls, auto bodies, etc. Several commercial types of glass bedding are available. Those having the least shrinkage are the more desirable. Some types have up to 7% shrinkage while others have none. Shrinkage does not ordinarily occur during the hardening or setting up process, but rather during the two or three day curing period following the hardening action. It is strongly recommended that a skilled gunsmith perform the bedding operation, which at first observation looks simple, but in reality calls for considerable skill, knowledge, and experience on the part of the craftsman.

d. Trigger Adjustment. Most high powered rifles of American manufacture have adjustable trigger mechanisms which can and should be adjusted to a relatively light pull of two to three pounds. In as much as the target may be visible for only a few seconds, it is very important that the counter-sniper be able to get off a fast shot with a minimum of disturbance to his rifle. Safety is also very important and the rifleman should practice frequently in order to become thoroughly accustomed to the sensitivity of the trigger so as to obtain a perfectly timed shot that is aimed only at his prescribed target and does not endanger innocent bystanders.

4. There is a common misconception among shooters that there is no longer a need to clean the bore of a rifle due to the non-corrosive priming used in the manufacture of modern ammunition. Nothing could be further from the truth. The bench rest shooters who are conscious only of accuracy, clean their barrels after every tenth record shot. The build up of metal fouling from the bullets passing through the barrel can greatly impair accuracy. The rifle should be properly cleaned after firing. Many types of bore cleaners are available, most of which will do a satisfactory job. The bolt should be removed and the rifle cleaned from the breech end, with care taken not to cause damage or wear to the rifling at the muzzle end of the barrel. Instructions and recommendations supplied with the cleaning agents should be followed as closely as possible.

5. There is one additional type of rifle that should be given consideration, and that is the .22 caliber rim fire. This rifle when equipped with a telescopic sight is very useful as a training rifle due to several factors previously mentioned such as cost, noise, recoil, etc. Equipped with a silencer and sub-sonic ammunition (some types of Match 22 long rifle cartridges are loaded below the speed of sound by the factory), it could be employed within a few yards of bystanders, etc., without their being aware that anything is going on. It is also helpful in eliminating certain undesirable domestic animals or wildlife without arousing the neighborhood with gun fire.

CHAPTER 3

SIGHTS

1. Let us now discuss one of the most important and yet least understood components of our countersniper equipment, the SIGHTS on the rifle.

Early firearms had no sights, but were simply pointed in the direction of the target, and a prayer to the right diety was considered to be very helpful. In fact, many firearms throughout history have had religious symbols placed on them to assure assistance from the gods when firing at infidels, non believers, and even certain species of game.

Later, front sights were installed, followed by open rear sights, peep sights, and finally the high quality and efficient optical sights which are available today. The modern telescope, permits one to see targets that are completely invisible to the naked eye; allows the shooter to utilize the full accuracy potential of the rifle and finally provides the countersniper with the confidence that he can hit his target, which is absolutely essential in activities of this sort.

2. In order to prove the requirement for an optical sight on the part of the countersniper, we shall make a comparison of the characteristics of all types of sights available today.

a. The Open Sight is found on most .22 caliber rim fire rifles and many hunting rifles when received from the factory. This type of sighting system requires the shooter to place his head and eye in an exact position in relation to the sight to insure that they are all properly aligned. He must focus on the rear sight, the front sight, and the target at various times in the firing of each shot. In as much as it is physically impossible for the human eye to focus on more than one of these objects at any one time, his eye must shift its focus from the rear sight, to the front sight, to the target. In the course of obtaining a good "sight picture" this process is actually run through several times in the firing of a single shot. Experience in competitive shooting has shown that the best results are obtained by firing with the front sight sharply in focus and allowing the rear sight and target to be out of focus. We immediately note a disadvantage to this system in the use of moving targets in as much as we tend to look directly at the target and consequently accept a loss of accuracy. Another objection to the open sight system is the fact that it covers up the target, to the degree that nothing below the top of the rear sight can be seen when aiming (see illustration). Additional targets, which certainly could be danger points, are obscured by the sights.

b. The peep sight is a much more precise aiming device than the open sight and requires the eye to focus only on the front sight and the target. Again, the best results are obtained while concentrating focusing on the front sight. However, we have the same problem with peep sights on a moving target as mentioned above, in that we may loose sight of it. The peep is simply looked through and the eye is supposed to naturally seek the brightest spot in the center of the peep. However, many people will tend to drift into the lower half of the aperature of the peep and consequently shoot low. The light passing through the aperature is much lower in intensity than the available light and many time objects that can be seen clearly with the naked eye cannot be defined by looking through the sights. The field of view which is defined as the area that can be seen through the peep sight, is limited. In fact, it is much smaller in size than the area that is viewed through telescopic sights of the magnification levels of which we are concerned (see illustration). Further, good peep sights are expensive and cost almost half as much as an acceptable optical sight which does the job much more effectively.

c. From experience it has been determined that the telescope sight is the only suitable sight to be employed in the countersniper role. In shooting, you simply look into the eyepiece of the telescope, place the reticle on the target and press the rigger.

3. We shall explain the terminology related to optical/telescope sights in some detail below in as much as the average person has little or no knowledge in this field:

a. RETICLE: This is the aiming device within the sight which is found in many configurations such as cross wires, post, center dot, or combinations of any and all of the above. The reticle, which is greatly magnified by the lenses of the telescope, is in reality an extremely small and delicate assembly that is not to be tampered with by the amateur. For example, the ordinary cross wires are manufactured from a material that is actually less than one half the thickness of a human hair.

70 YARDS DISTANCE

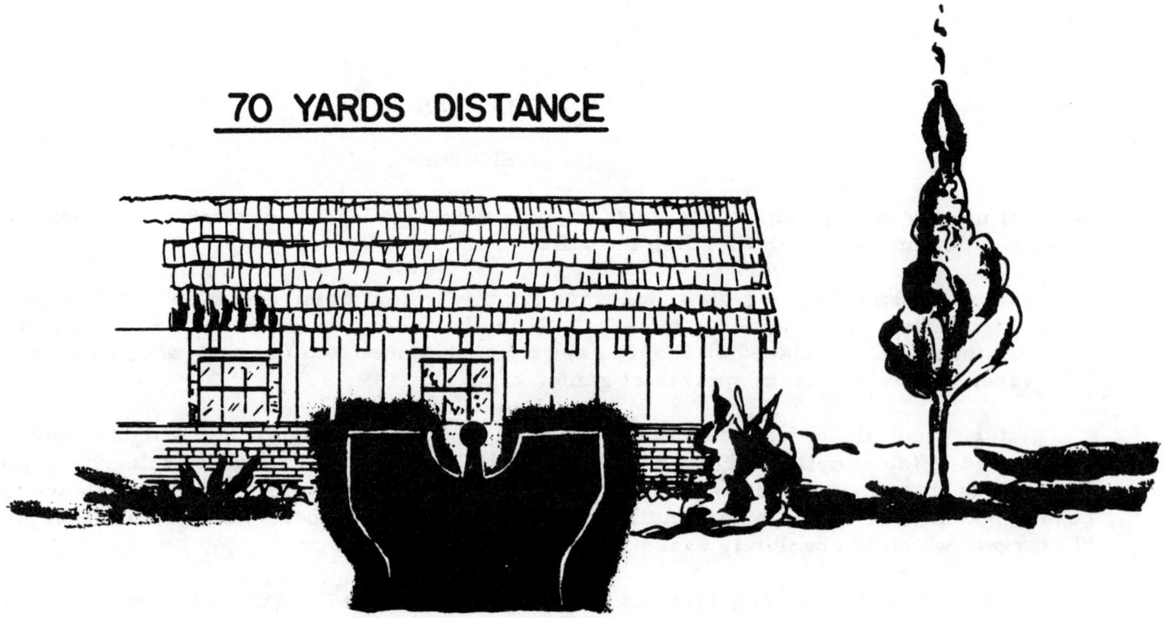

VIEW AS SEEN WITH FACTORY OPEN SIGHTS

(1)

70 YARDS DISTANCE

VIEW AS SEEN WITH PEEP SIGHTS (M16 RIFLE)

(2)

SIGHT PICTURE WITH PEEP SIGHTS (M16 RIFLE)

VIEW AS SEEN WITH 6 POWER SCOPE AND RETICLE SUITABLE FOR ALL TYPES OF LIGHT CONDITIONS.
(3)

b. FIELD OF VIEW: This is the "picture" that one sees through the eyepiece of the telescope. Its size is sometimes described in degrees of angle, but more commonly expressed in terms of a certain number of feet in width at 100 yards. In looking into the eyepiece you can see a framed picture of the target with the reticle on the same focal plane. In effect, the result is similar to drawing a set of cross wires on a photograph. Unlike the metallic sights which requires one to shift his focus from target to the sights, the telescope presents everything sharply in focus to the viewer.

c. EYE RELIEF: Is defined as the distance of the eye ırom the telescope eyepiece which allows the viewer to obtain a maximum size "picture." In the average low power, hunting type telescope, the correct distance falls between 3 to 5 inches from the rear of the telescope. However, the eye relief in target type scopes may be as short a 1 1/2 inches. Heavy recoiling rifles require long eye relief to prevent the shooter from being struck in the face by the rear of the scope on full recoil. It is not necessary to look into the center of the eyepiece when using the telescopic sight. If you can see your target and the center of reticle you can shoot. With metallic sights the position of the head and eye is extremely important, but with the telescope, this is relatively unimportant.

d. EXIT PUPIL: If you hold the scope out at arms length, you will see a small circle of light in the eyepiece. This spot is usually about one quarter of an inch in diameter. Its size can be utilized when comparing scopes for use in unfavorable light conditions. Generally speaking, the larger the exit pupil, the better one will be able to see at night.

e. Parallax: This is a term that shooters hear of but seldom understand. When a shooter has trouble with the zeroing or grouping characteristics of his rifle, some friend will ask, "How's the parallax?" Not wishing to display his ignorance, he will invariably reply, "It's OK," when in reality he doesn't know what it's all about. When you look into the eyepiece of a telescope, the reticle and image should appear on the same focal plane. If the scope has plain cross wires, the wires should appear as though they were drawn on a photograph and should not require a shifting of the focus of the eye to see each one separately.

To check for parallax, set the rifle with scope (or scope alone) in some sort of a solid rest, i.e. sandbags, vise, etc., and aim at a target at 150 to 200 yards. Without touching the rifle or scope, move the head about, so that the eye moves in the area of the exit pupil. There should be no movement of the reticle on the target, as you are moving only your eye and not the rifle. If there is movement, the telescope is not properly adjusted for parallax. In scopes with adjustable front lenses, the parallax can be corrected by following the manufacturer's instructions. In almost all cases, defective hunting type telescopes should be sent back to the factory or at least returned to the maker's local representative for parallax adjustment.

Parallax can be demonstrated in a simple way at your desk. Use a photo that has some sort of suitable target on it. Place the tip of an ordinary pencil on the target with the pencil in a normal writing position. Move your head about; there will be no movement of the pencil point on the photo. Now raise the tip of the pencil a couple of inches off of the photo with the hand still resting on the desk top. Now move the head and notice the relative movement of the pencil point on the photo. This is comparable to the error in the telescope, when not properly adjusted for parallax which means simply that the reticle and picture are on different focal planes.

What does all this add up to? It simply means that if your eye is looking into the right side of the eyepiece you shoot on one side of the target and with your eye on the left you shoot on the other side. The center of impact is changed by the position of the eye when using a telescope with parallax error, but it is not influenced by eye movement on a correctly adjusted telescope.

4. Inasmuch as the countersniper normally becomes involved in actions in other than favorable light conditions, there are several areas of concern in relation to the type of telescope sight that is required by the countersniper.

a. The magnification of the telescope influences the amount of light that can be seen in order to identify a target. The low power scopes have a bright image; the high powered scopes, a dim one. The six power telescope is recommended as the happy medium for the .222 caliber. Its light gathering qualities are very similar to the so called "Night Glass," the 7 x 50 bioncular, which has proven itself for use under unfavorable light conditions world wide. We are referring to the hunting type of sight with internal adjustments for windage and elevation, which will remain at a constant range setting and utilize the trajectory curve of the cartridge, by using a simple hold over or under technique within the 300 yards usable range.

For the 30 caliber longer range rifles, a target-type telescope of ten power is recommended, which must have precise micrometer adjustments to obtain exact zero at these greater ranges. The scope will not be usable at night except under ideal conditions. However, there is really no high-power scope that is completely suitable for night use.

b. The reticle is our next concern. Remember that it is not enough just to be able to identify a target, one must also be able to take a precise aim at the target. Restrictions on night hunting throughout most of America have placed little or no demand on our telescope sight makers to stock a reticle suitable for night shooting. Fine, medium, or even heavy cross wires are invisible at night, the center dot is almost impossible to find, and the post too narrow to be seen against our target.

Two pages of reticle types are shown. The first six (A through F), are those considered to be satisfactory for all light conditions. The second six (G through L) are considered borderline or unsatisfactory. Each one will be discussed as to its merits or limitations.

5. The telescope sight is no better than its mounting system. The sight must be mounted permanently and as ruggedly as possible. Once the shooter has become familiar with the telescope and is knowledgeable of its capabilities, he will no longer consider matallic sights. In order to ensure that the student rapidly assumes this attitude, and in order to accelerate training, the metallic sights should be removed from the rifle. Any job that the countersniper may be called upon to do can be accomplished much more effectively with the proper telescopic sight than with any other type of sight.

Personnel must be trained to maintain their equipment and to recognize any looseness or movement in the sighting system. They must be aware of the fact that the telescope is mounted on a short radius. The mounting rings are only four or five inches apart normally, so that a few thousandths of an inch of play can cause a large error at the target. Insofar as rifles with metallic sights are concerned, the radius (distance between front and rear sights) is about thirty inches, so that a thousandth of an inch or two error at one end will not result in a significant error at the target.

In the case of the hunting-type scope, the mount which forms a one piece bridge across the top of the receiver is the most desirable, for it also assists in supporting the heavy free floating barrel. This bridge type base is particularly desirable if you are using using the Model 98 Mauser action with the deep thumb cut in the left receiver wall. (See illustration on page 17.)

The base or bases, (depending on the type of scope), whether it be a hunting or target type, should have the screws pulled up extremely tight and some type of locking compound used to prevent them from vibrating loose. The most rugged telescope mount ever manufactured has its weakness, so be forewarned and check all screws before trouble develops.

RETICLE TYPES

RETICLES CONSIDERED SATISFACTORY FOR USE

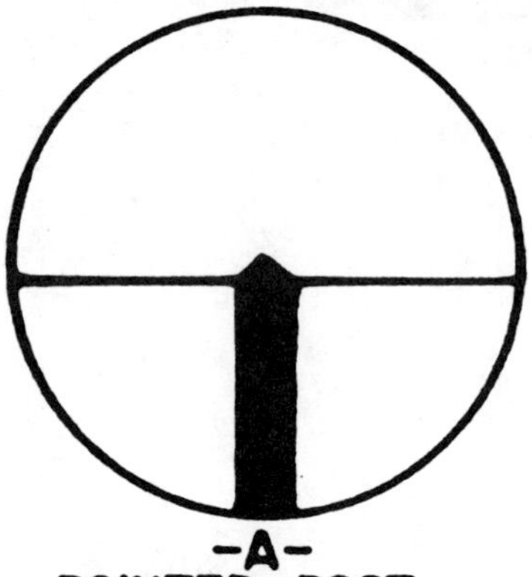

-A-
POINTED POST
WITH HORIZONTAL WIRE

-B-
POINTED POST
WITH SIDE POSTS

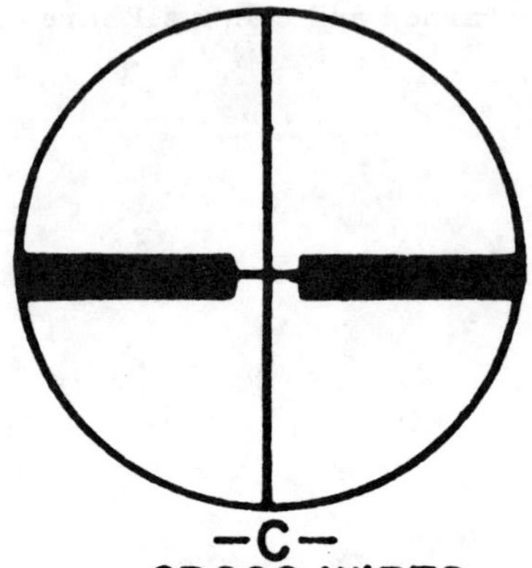

-C-
CROSS WIRES
WITH SIDE POSTS

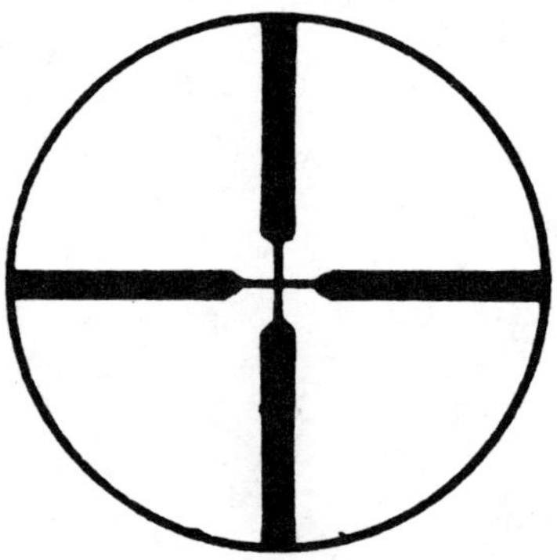

-D-
MULTIPLE POSTS
WITH CENTER CROSSWIRES

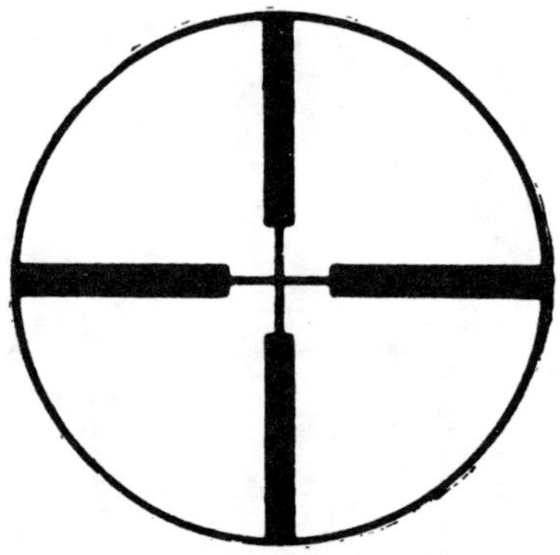

-E-
MULTIPLE POSTS
WITH
CENTER CROSSWIRES

-F-
POINTED POST

SATISFACTORY RETICLES:

Figure A

This is probably one of the best reticles available, but unfortunately it is seldom found in the United States. Note that the post is parallel sided and that the point has a 90° included angle. The point of the post can be seen against any type of background and a precise aim can be taken with it. There is no confusion as to the cross wire representing an impact point. The parallel sides of the post and the horizontal wire cause the shooter to keep his rifle verticle and eliminate the errors caused by "canting" the rifle to the right or left.

Figure B

This is very similar to "A", and affords just as good an aim, but the side posts, which assist in night shooting, tend to clutter the field in daylight. It is the "Standard" reticle used in Europe, where considerable night hunting is done.

Figure C

This is an excellent reticle, allowing precise aim under favorable conditions. At night the target is simply placed between the ends of the side posts. The aim, however, would not be as precise as that afforded by the pointed vertical post.

Figure D and E

These reticles are available in slightly different configuration from several American manufacturers. Of the two shown, one has pointed ends on the posts, the other with square ends; either is suitable for night and day use.

Figure F

This has the same post with pointed end as found in "A" and "B" but the lack of some type of horizontal line makes it appear empty. You have a feeling that something is missing. It allows precise night and day aiming.

RETICLES CONSIDERED UNSATISFACTORY

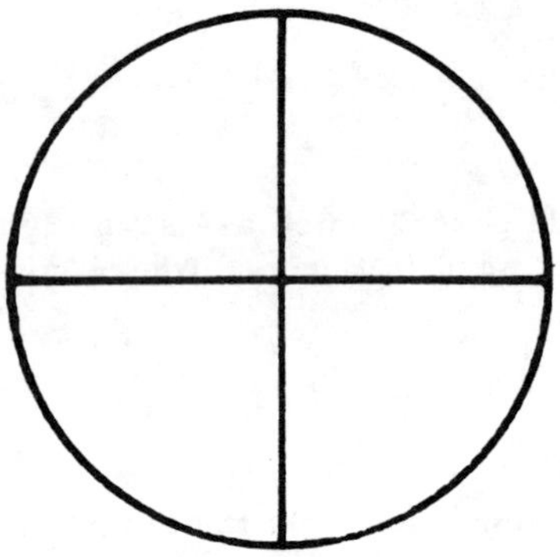

-G-
CROSSWIRES

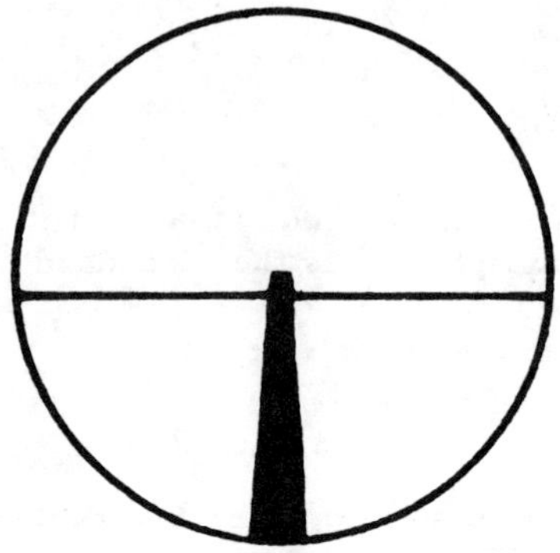

-H-
TAPER FLAT TOP POST WITH HORIZONTAL WIRE

-I-
FLAT TOP POST

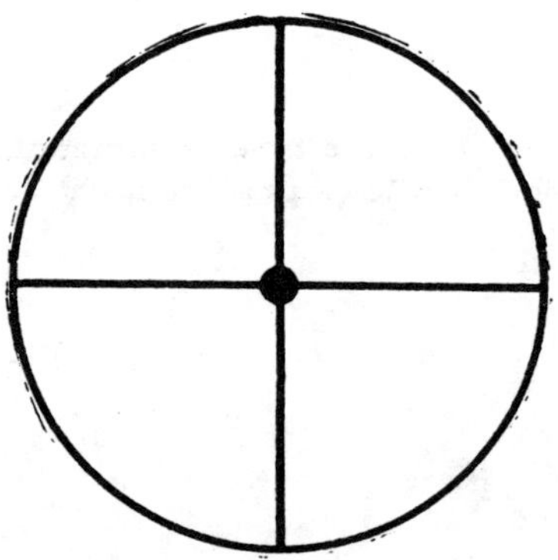

-J-
CROSSWIRES WITH DOT

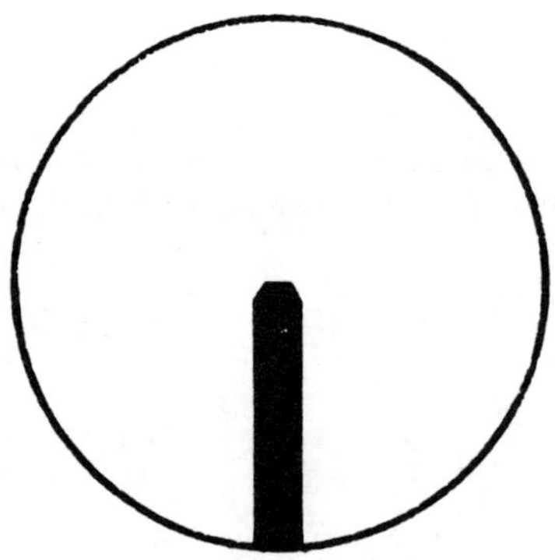

-K-
BLUNTED POST

-L-
FINE POINTED POST

UNSATISFACTORY RETICLES:

Figure G

This is the most common reticle found in the United States. However, although it is suitable for use in bright light, it is absolutely ineffective at night.

Figure H

The second most common reticle found in the United States. This is actually a very poor reticle for practically any purpose. The flat top does not allow a precise aim and is too narrow for night use. The tapered sides of the post tend to cause the shooter to "cant" the rifle by paralleling one side or the other with some vertical object in the background. A combination of a narrow post and a horizontal wire may cause an excited shooter to use this intersection as an aiming point rather than the top of the post.

Figure I

The flat top of this reticle does not allow a precise aim, therefore it is not considered satisfactory for sniper use.

Figure J

Although an excellent reticle for use in good light, it becomes lost in the background under unfavorable light conditions.

Figure K

This reticle is partially effective at night, but does not allow the precise aim of the reticle with the 90° angle point.

Figure L

The 60° angle of this post allows for precise aim in good light, but the point fades out completely at night, resulting in large errors in the elevation of the impact of the bullet.

CHAPTER 4

NOISE AND MUZZLE FLASH

1. In the confusion of a riot torn area, there will be times when the presence of a well-trained rifleman equipped with a telescope sighted rifle will have a sobering effect on highly excited people. Troops equipped with automatic weapons, such as the Thompson submachine gun, also produce the same effect. However, in the accomplishment of most missions, it is desirable that the countersniper be as inconspicious as possible; therefore, two aspects of rifle firing must be considered: noise and muzzle flash.

2. The sound produced by a high powered rifle is certainly going to be heard above the normal sounds of the city, particularly at night in built up areas. There are several ways to reduce the sound of a firearm, i.e., shooting through a pillow, clothing, or a box of wadded paper, etc. None of these systems will permit the rifleman to aim his shot and most certainly would influence the accuracy potential of the rifle. One technique is to fire from the back side of a room and to shoot through a partially opened window. However, this limits the rifleman's field of fire to a very small area. The most efficient technique available to date is the use of a noise moderator, commonly referred to as a "Silencer." The "James Bond" type of silencer does not exist, except in the movies. The development of satisfactory moderators is very difficult to say the least. It is pointed out that the expanding gases from the fired shot are very difficult to contain because of the large volume and high pressure level at the muzzle which varies with the caliber of the round. A large but properly designed silencer has been developed for the 22 caliber rim fire rifle, which was possible due to the relatively small volume of gases produced by the round. This is the most efficient moderator produced to date. Only the functioning of the firearm mechanism can be heard when the better moderators are employed. The high power rifle is another matter. The moderators developed to date for the high power rifle are much less efficient than those developed for rim fire cartridges. The volume of gases produced is so great that the gases can only be slowed and released over a longer period of time. The best moderators presently in production for high power rifles reduce the noise to about 25% of the normal level which aids in concealing the sound source, thereby making detection of the sniper's location more difficult at the greater ranges.

3. The bolt action mechanism lends itself very well to the use of a silencer, since all the gases are moving toward the muzzle. The automatic or semi-automatic rifle presents problems, however, as a silencer is designed to hold the gases and to release them slowly (relatively speaking). In a self loading mechanism, these gases try to go in both directions as the breech mechanism is opened and residual gases along with powder and primer residue tend to come back through the barrel and strike the shooter in the face. Special silencers are made for the automatics which "Bleed Off" part of the gases in another direction, so they will not incapacitate the rifleman by bringing tears to his eyes on subsequent shots.

4. A properly designed and installed silencer is not a detriment to accuracy. Rather, the rifleman generally performs much better with a silencer installed on his rifle because of both physical and psychological factors, such as, the reduction in recoil due to the muzzle brake effect of the silencer baffles, and the fact that the rifleman feels more secure in that he has less chance of being located due to the reduction in noise.

5. An improvement in the grouping characteristics of many rifles is noted after the installation of a silencer, which is due in part to the dampening effect that the weight of the silencer has on barrel vibrations and to better stabilization of the bullet as it leaves the end of the silencer, because it is traveling through a much lower "exit pressure" than from the muzzle of an unsilenced rifle. To explain this function, it should be noted that no bullet is absolutely perfect and as a bullet leaves the barrel, the gases try to get around it, causing a deflection in relation to the irregularities in the base of the bullet. The pressure level of the gases involved at the exit determines the magnitude of the deflection. So, the lower the pressure, the less deflection, and the better the accuracy, all other factors being equal.

6. The silencer is also an excellent flash suppressor. Although the muzzle flash may not be of any importance in daylight, it can be a dead give away at night. (We know of no flash suppressor available on the commercial market; however, a competent gunsmith can fit the one used on the M-14 to practically any rifle.)

7. To be used effectively, the rifle must be sighted in with the silencer properly installed. The zero of the rifle must be reestablished each time that the silencer is removed and replaced, otherwise

the rifleman has no guarantee that he will hit his target. Properly made silencers are very expensive. The personnel that install them on the rifle must be extremely careful to assure that the hole through the silencer is perfectly aligned with the bore of the rifle. It is not only embarrassing, but downright dangerous for a bullet to deflect and emerge through the side of the silencer. Caution: An alignment gauge should be used each time the silencer is installed. "Do not place any trust in the fact that it has always been 'lined up' in the past."

CHAPTER 5

GENERAL NOTES

1. Many books have been written about the efficient use of firearms. Most deal with the "classical" positions used by a competive shooter, and consequently will not be referred to. The countersniper is a hunter and must use any and all tricks of the trade to assure a proper hit. The lives of his fellow officers and that of the general public are at stake. Time is extremely critical, therefore, he can expect to be required to make shots at varying angles and distances on a split second's notice. The hunting of varmints such as woodchucks and crows provides outstanding training because the techniques involved are almost identical. This sort of training may be impractical in most areas.

2. The illustrations provided in Annex C demonstrate the proper way to support a weapon. Note that the hands holding the rifle are themselves resting on solid support. It is recommended that the shooter obtain a good fitting pair of buckskin gloves, which will allow him to take a comfortable position on rough, hot, or cold surfaces, thereby increasing his effectiveness. It is advisable to learn to use the gloves at all times in shooting, so that the critical shot will be delivered correctly to its mark in a timely and confident manner.

3. The light trigger pull previously recommended will be strange to many who are not familiar to a finely adjusted trigger. With training, the shooter will be able to time his shot to the exact fraction of a second which is often required when engaging an elusive target. "Dry firing" on an empty chamber while aiming at small objects within the range capability of the rifle is strongly recommended. It costs nothing but time.

4. Aiming with the telescope sight is so simple that there may be a tendency to become over confident if you shoot only at motionless targets. An excellent method for becoming familiar with the pointing qualities of your rifle is to stand inside a building, aiming and dry firing at the hub caps of passing cars. The secret of using the scope is to look at the target and simply place the scope between your eye and the target, not to look into the scope and then try to find your target. You know the location of the target, so again, keep your eyes on the target and simply place the scope in between your eye and the target. That is all there is to it. Practice in the handling and pointing of your rifle will allow one to place his shots within a few inches of the center of the target. Again, let us remind you that it is not necessary to look into the center of the eyepiece. As long as you can see your target and reticle, the rifle will be pointed at the aiming point of the reticle.

5. Two illustrations are presented in Annex D which depict typical countersniper employment situations. Some basic rules should be noted:

a. In the majority of situations, the countersniper should be in an elevated position, in order to give him better observation of the area.

b. Security and communications for the countersniper should be provided by an assistant (observer or supervisor) who should have some means of communication with higher authority.

c. The sniper team is in an excellent position to provide information to the controlling authority.

d. To assist in the identification of targets and gathering of information, the observer should be provided with good quality binoculars.

e. The presence of a partner also has a stabilizing effect on the rifleman in that he is not "alone" but can depend on his partner for advice and morale backing.

6. It is possible to proceed giving advice on thousands of situations and ideas, but the first shot the countersniper is called on to deliver will most likely involve an entirely different situation to any that may be presented. No two situations are ever the same, inasmuch as the ingredients involved always vary. Therefore, we shall continue no further, other than to note that the most important ingredient of any operation is good clear thinking on the part of the countersniper. There is no substitute for good common sense. On that we end.

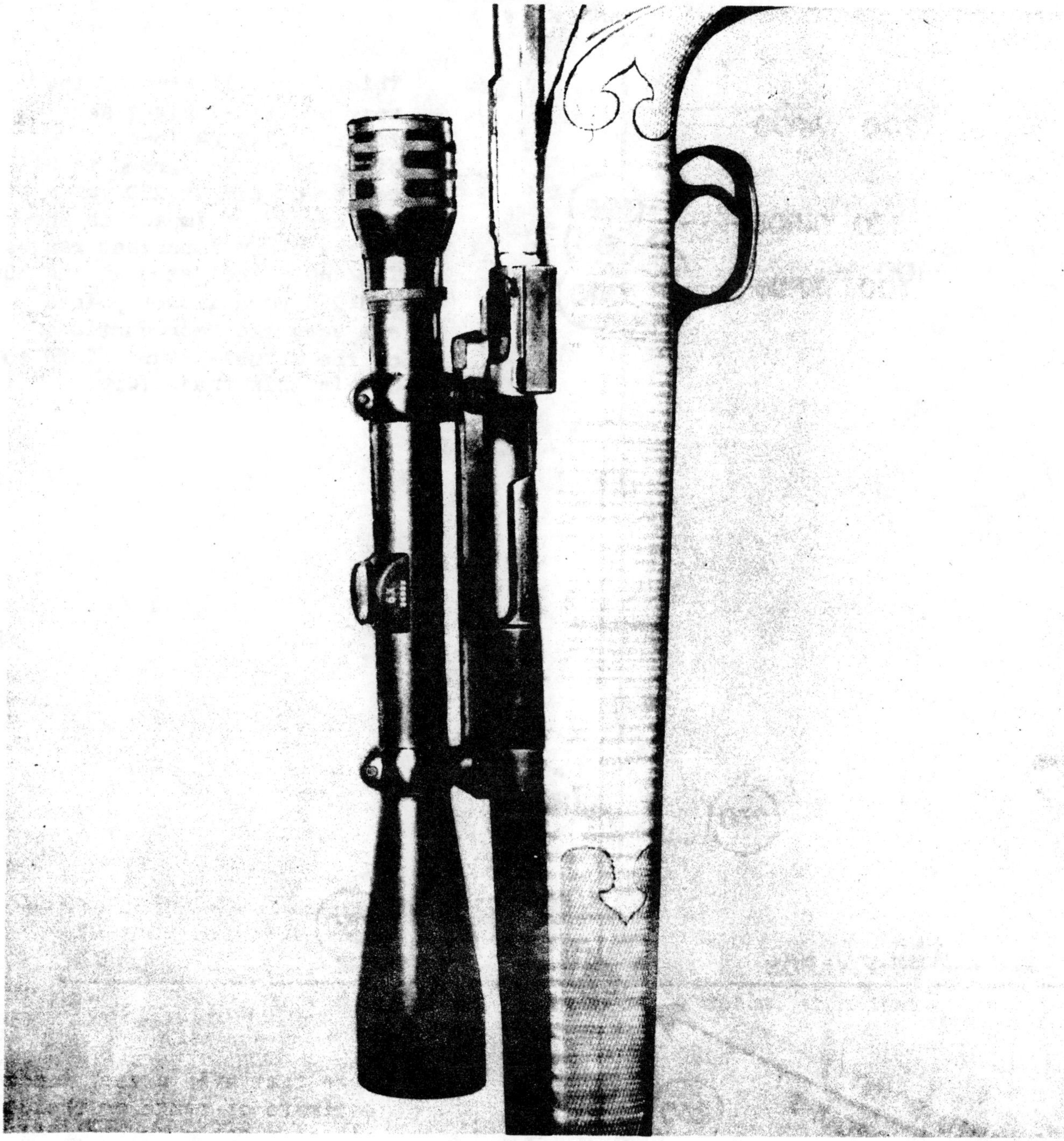

ONE PIECE BRIDGE MOUNT WITH SCOPE

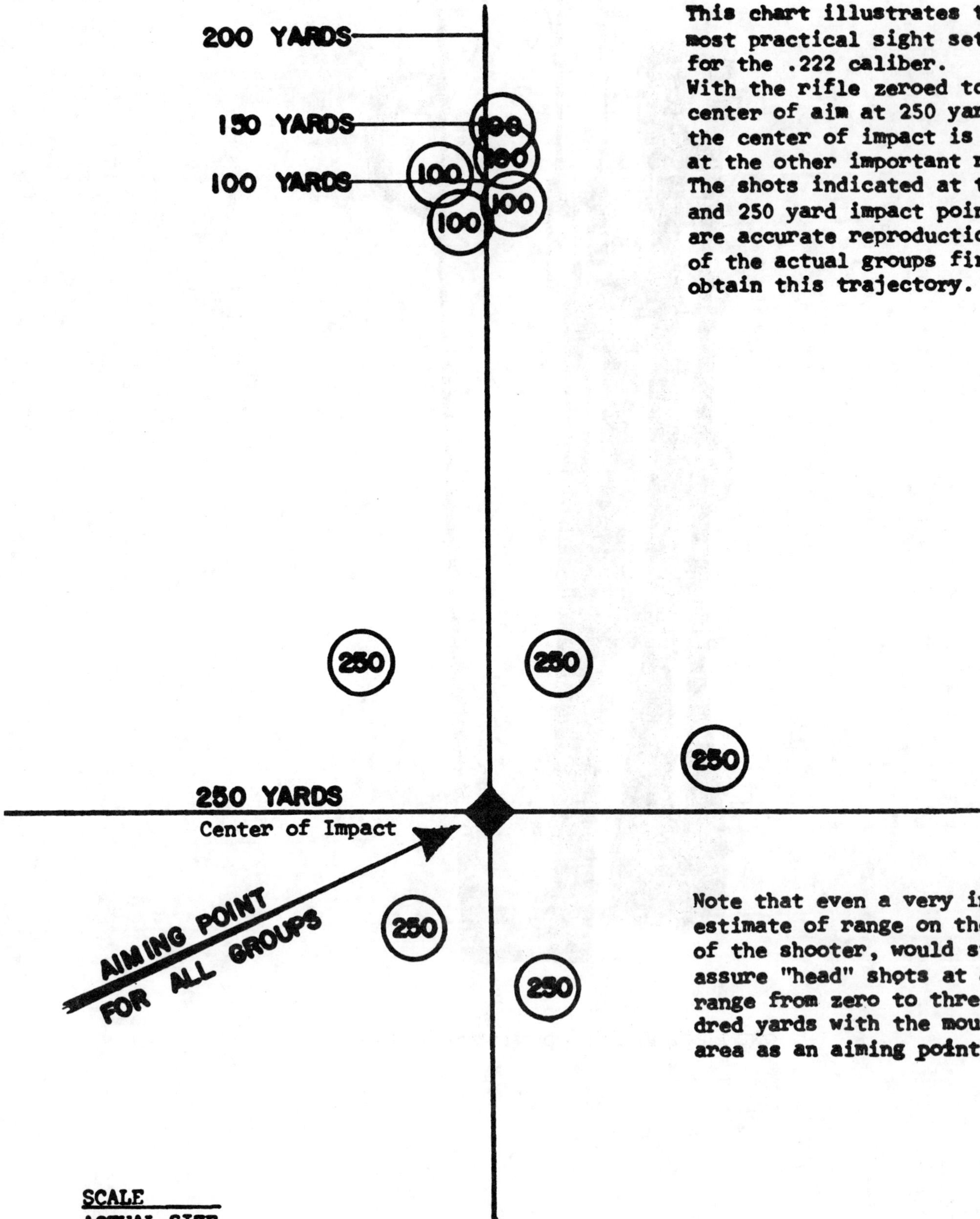

This chart illustrates the most practical sight setting for the .222 caliber. With the rifle zeroed to hit center of aim at 250 yards, the center of impact is shown at the other important ranges. The shots indicated at the 100 and 250 yard impact points are accurate reproductions of the actual groups fired to obtain this trajectory.

Note that even a very inaccurate estimate of range on the part of the shooter, would still assure "head" shots at any range from zero to three hundred yards with the mouth/nose area as an aiming point.

SCALE
ACTUAL SIZE

With the rifle sighted in for 250 yards, the 300 yard impact point would be centered in the area shown using factory loaded .222 Remington Cartridge.

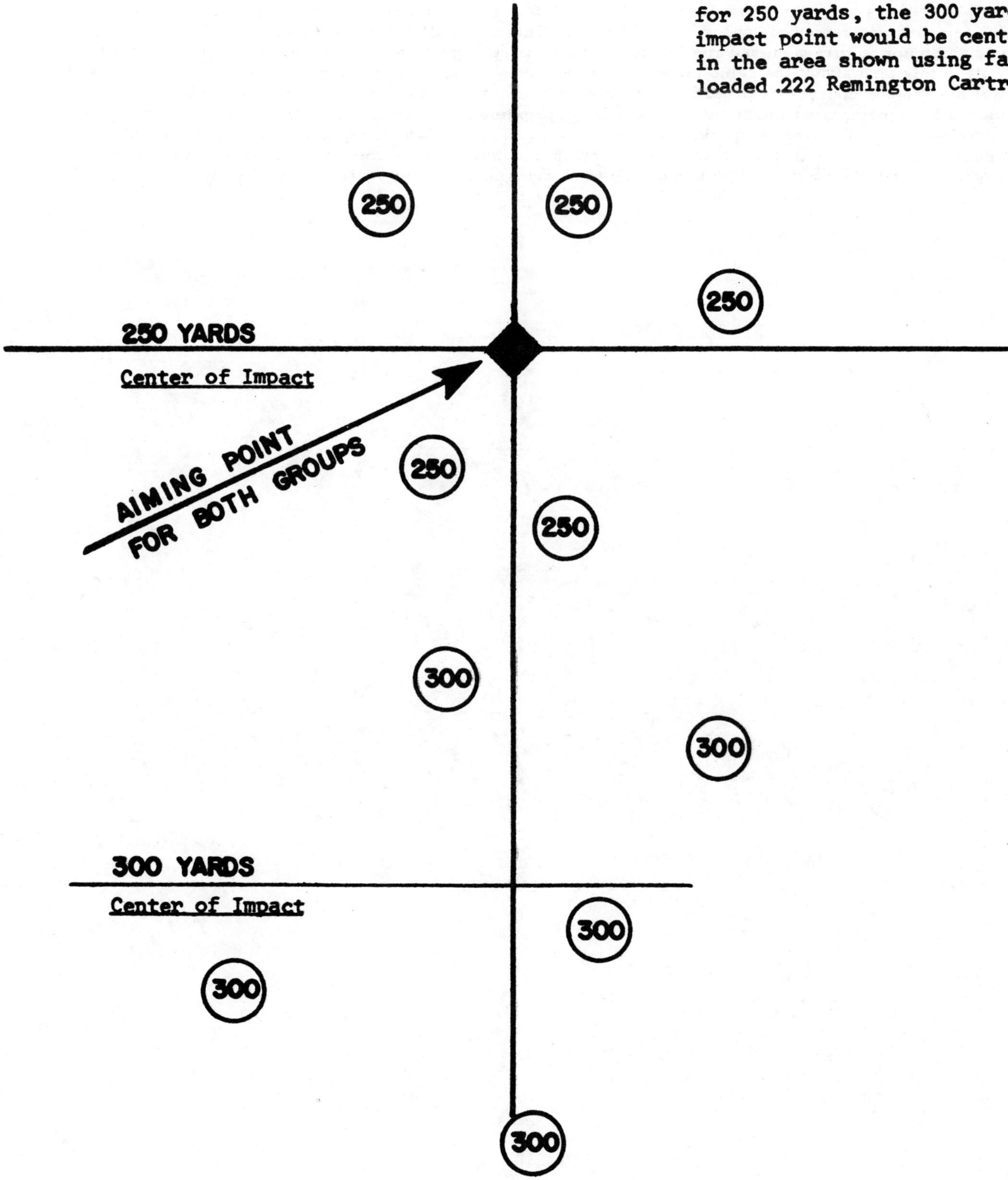

SCALE
ACTUAL SIZE

ANNEX B

Epoxy Impregnation Procedures.

The stocks are placed in a large container (8 feet in diameter), the lid is toggle bolted down, and the temperature raised to approximately 300°F. This turns all moisture in the stock to steam. A vacuum pump is turned on and run for about one hour, this removes just about all moisture. While still in this tank and at the same temperature, an epoxy is run in, in a liquid state, and held at 100 PSI for another hour. The pressure is then slowly lowered and the stocks removed. They are then placed in a curing oven where they remain for approximately three days. This is to set up the epoxy and also to fill all of the sap pockets and pores with an epoxy, which replaces the water. This procedure increases the tensile strength of the stock and completely eliminates warpage. Any further expansion and contraction of the stock, due to changes in moisture content is negligible.

Unstable Position

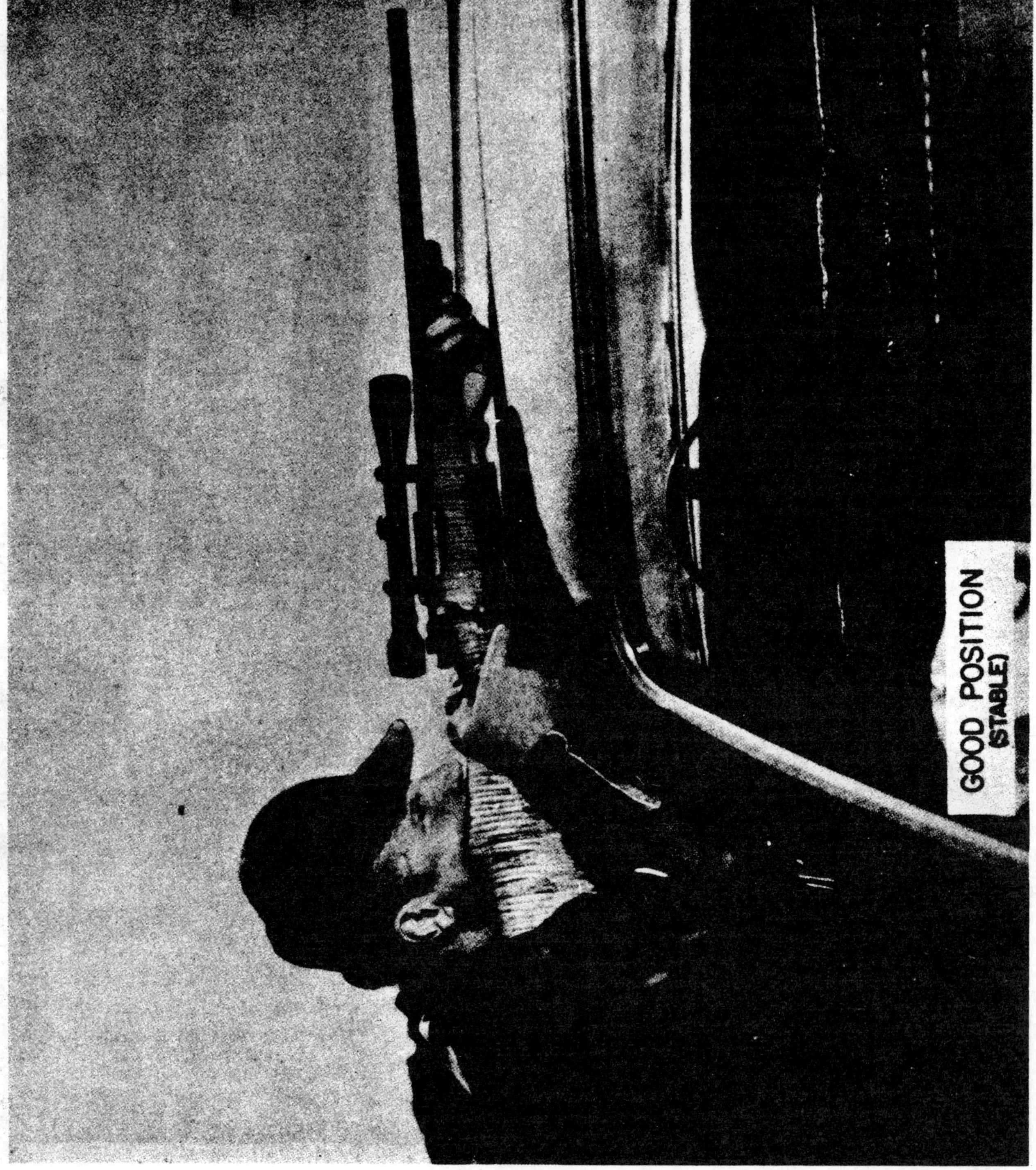

Stable Position

ANNEX "C" #3

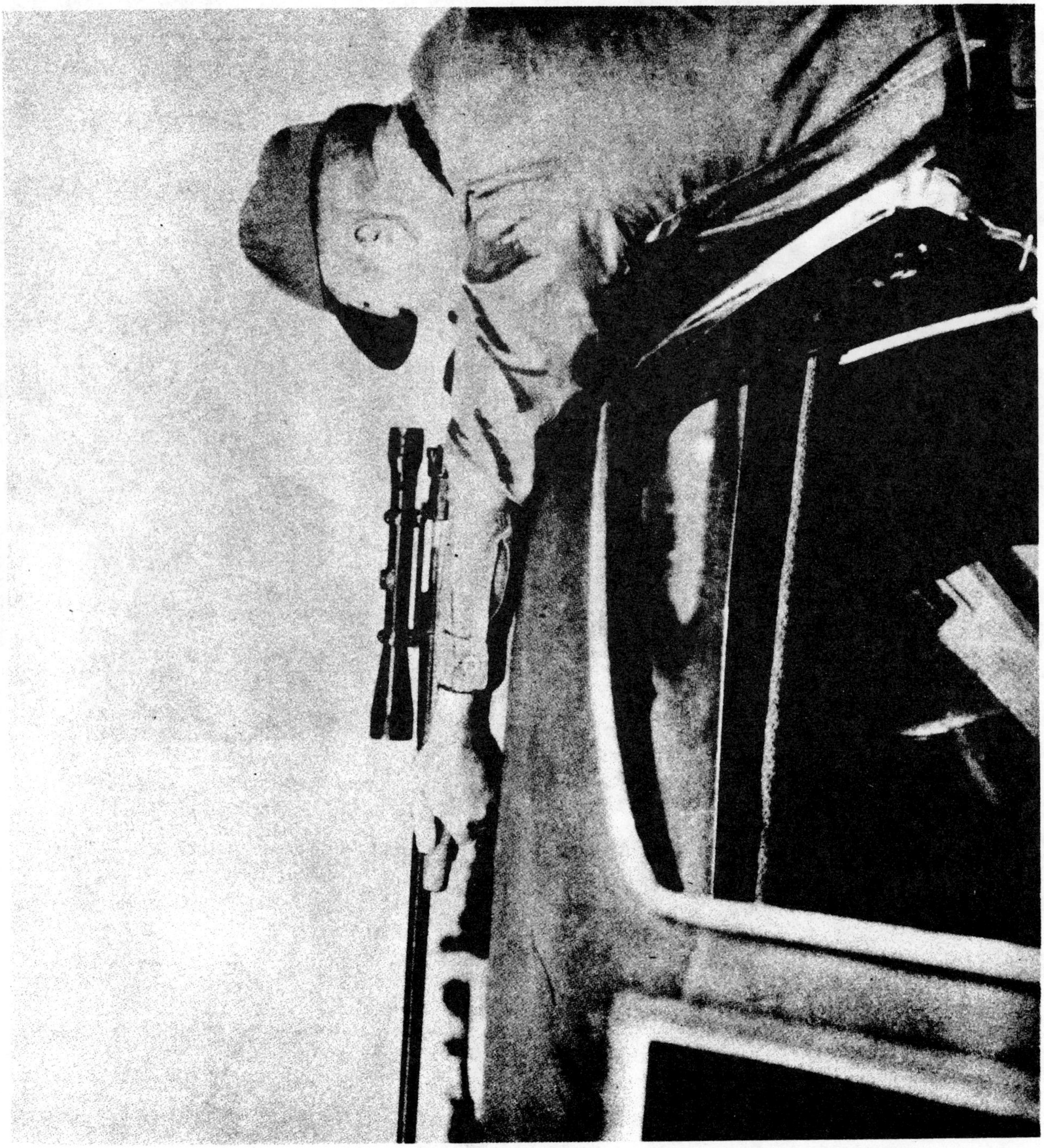

Stable Position

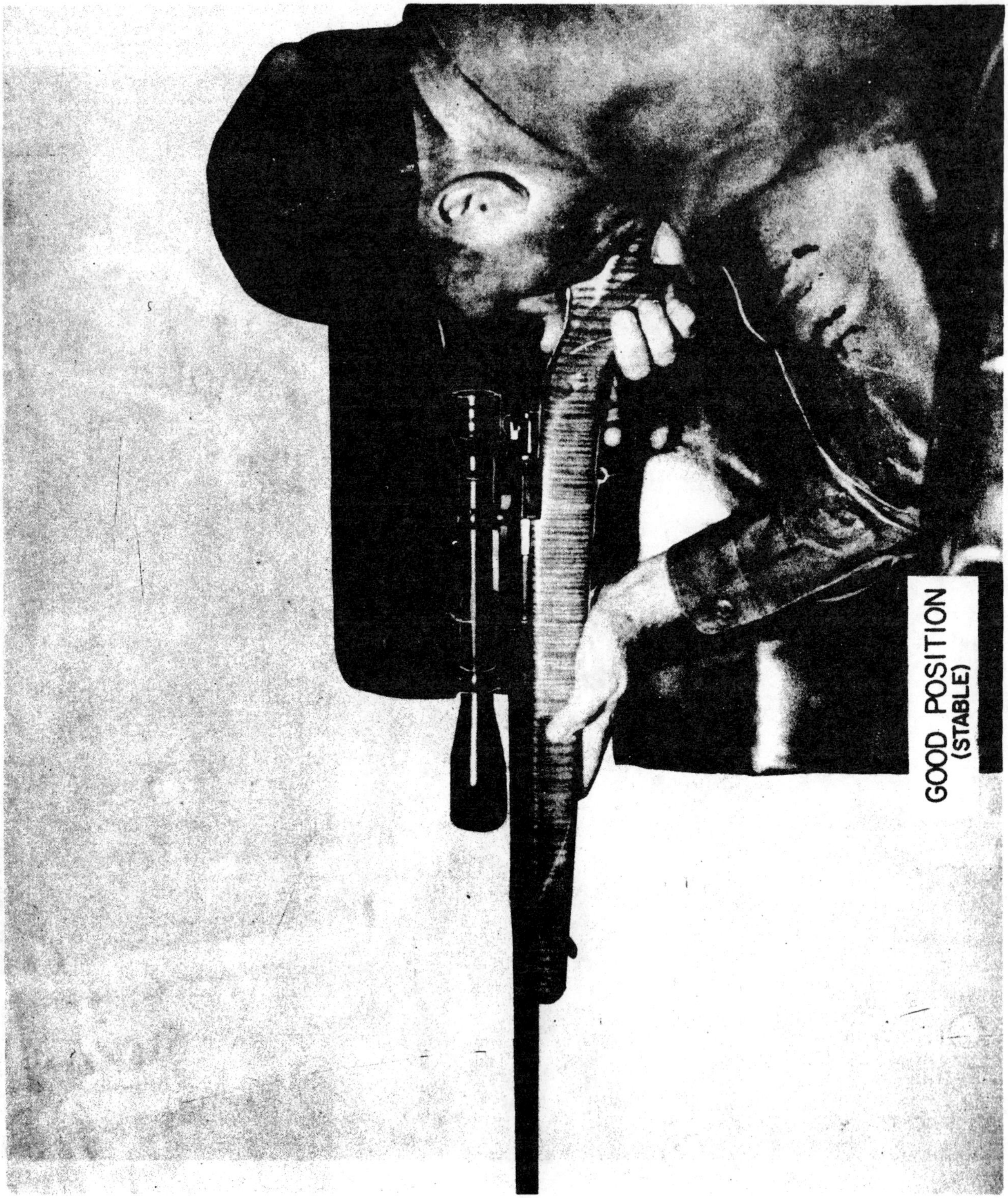

Stable Position

ANNEX "D" #1

Typical Countersniper Situation

ANNEX "D" #2

Typical Countersniper Situation

ANNEX E

THIS IS A LIST OF ITEMS CONSIDERED SUITABLE FOR COUNTERSNIPING.

ITEMS ARE LISTED IN ORDER OF PREFERENCE
THIS DOES NOT CONSTITUTE AN ENDORSEMENT OF ANY
PRODUCT OR CONDEMN THOSE WITH WHICH WE ARE NOT FAMILIAR

Telescopes:

Redfield	- 6X with 4 post cross hair reticle
Leupold	- 7.5 X with duplex reticle
Unertl	- 6X Condor with post reticle
Realist	- 6X Camputer with standard reticle
Tasco	- 6X Super Deerslayer with appropriate reticle

Scope Mounts:

Redfield	- Brigde mount with 1" rings
Buehler	- Bridge mount with 1" rings
Weaver	- Top detachable 1" rings

Rifles:
- Remington model 700 Varmit Special
- Winchester model 70 Target Heavy Barrel
- Remington model 40XB Target Heavy Barrel

Silencers and silenced rifles: - Sionics Inc., Rt 1, Powder Springs, Georgia

